INTEGRATED REVIEW WORKSHEETS

LISA YOCCO
Georgia Southern University

COLLEGE ALGEBRA IN CONTEXT WITH INTEGRATED REVIEW

Ronald Harshbarger
University of South Carolina – Beaufort

Lisa Yocco
Georgia Southern University

PEARSON

Boston   Columbus   Indianapolis   New York   San Francisco
Amsterdam   Cape Town   Dubai   London   Madrid   Milan   Munich   Paris   Montréal   Toronto
Delhi   Mexico City   São Paulo   Sydney   Hong Kong   Seoul   Singapore   Taipei   Tokyo

The author and publisher of this book have used their best efforts in preparing this book. These efforts include the development, research, and testing of the theories and programs to determine their effectiveness. The author and publisher make no warranty of any kind, expressed or implied, with regard to these programs or the documentation contained in this book. The author and publisher shall not be liable in any event for incidental or consequential damages in connection with, or arising out of, the furnishing, performance, or use of these programs.

Reproduced by Pearson from electronic files supplied by the author.

ISBN-13: 978-0-13-404022-6
ISBN-10: 0-13-404022-8

4

www.pearsonhighered.com

Table of Contents and Review Objectives for *College Algebra in Context with Applications for the Managerial, Life, and Social Sciences*, Harshbarger/Yocco

College Algebra in Context
Integrated Review Worksheets
Chapter 1

Objective 1. Write sets of numbers using descriptions or elements

Main Idea

A **set** is a well-defined collection of objects, including, but not limited to, numbers. One way to define a set is to list the elements (or members) between braces. For example, we may say the set A contains the elements 3, 5, 7, -8, and write $A = \{3, 5, 7, -8\}$. To say 3 is an element of A, we write $3 \in A$. Another way to define a set is to give its description using set-builder notation. For example, the set of all integers between 1 and 8, inclusive, can be written $\{x \mid x \text{ is an integer between 1 and 8, inclusive}\}$.

The set $N = \{1, 2, 3, 4, 5, \ldots\}$ is called the set of **natural numbers** (sometimes called the counting numbers)

The set $W = \{0, 1, 2, 3, 4, 5, \ldots\}$ is called the set of **whole numbers**.

The set $I = \{\ldots, -5, -4, -3, -2, -1, 0, 1, 2, 3, 4, 5, \ldots\}$ is called the set of **integers**.

The set with no elements is called the **empty set** and is denoted $\varnothing$.

Exercises

1. Write "the set of all natural numbers less than 8" another way.

_____

2. Write "the set A containing all natural numbers greater than 10" another way.

_____

3. Is it true that $3 \in \{1, 3, 5, 7, 9\}$?

_____

4. Write the set $\{1, 3, 5, 7, 9, 11\}$ as a description.

_____

5. Write the set $\{x \mid x$ is an odd number divisible by 2$\}$ in another way.

_____

6. List all the elements of the set $\{x \mid x$ is an even whole number between 9 and 21$\}$.

_____

Objective 2. Identify sets of real numbers as being integers, rational numbers, and/or irrational numbers

Main Idea

A **real number** is any number that represents a quantity along the number line. Real numbers can be rational or irrational. **Rational numbers** include integers, fractions containing only integers (with no 0 in a denominator), and decimals that either terminate or repeat. Irrational numbers are real numbers that are not rational. Examples are:

Rational: $-9, -9, \dfrac{2}{3}, 12, -\dfrac{10}{7}, 4.87, 3.\overline{45}$

Irrational: $\pi, \sqrt{2}, \sqrt[3]{-10}$

---

7. Is it true that $\dfrac{1}{2} \in N$ if N is the set of natural numbers?

_____

In Exercises 8-10, identify the sets of numbers as containing one or more of the following: integers, rational numbers, and/or irrational numbers.

8. $\{5, 2, 8, -6\}$

_____

9. $\left\{-1, \dfrac{3}{4}, 2.\overline{3}, -2\dfrac{5}{7}\right\}$

_____

10. $\left\{ \sqrt{5},\ \pi,\ \sqrt[3]{11},\ -\dfrac{\sqrt{3}}{8} \right\}$

---

11. Consider the set of numbers $\left\{ -2,\ 0,\ 3.8,\ 2\pi,\ \sqrt{6},\ 5.\overline{7},\ \sqrt{9} \right\}$.

List the numbers in the set that are

a. Natural numbers

b. Integers

c. Whole numbers

c. Rational numbers

d. Irrational numbers

e. Real numbers

---

12. Write examples of two numbers that are rational; two numbers that are irrational.

Objective 3. Graph numbers on a number line

Main Idea

We can represent real numbers on a number line. Exactly one real number is associated with each point on the line, and we say three is a one-to-one correspondence between the real numbers and the number line. Notice the rational numbers $-4.\overline{3}$ and 6.58 and the irrational numbers $\pi \approx 3.14$ and $\sqrt{3} \approx 1.732$ on the real number line below.

13. Graph the following real numbers on a number line, and state whether the number is rational or irrational.

a. -6

b. 4.75

c. $-2.\overline{3}$

d. $\sqrt{5}$

e. $\dfrac{12}{5}$

Objective 4. Work with set operations

Main Idea

Two sets are equal if they contain exactly the same elements.

Set A is called a **subset** of set B if each element of A is an element of B. This is denoted $A \subseteq B$.

If sets C and D have no elements in common, they are called **disjoint**.

The set containing the elements that are common to two sets is said to be the **intersection** of the two sets. The intersection of A and B is written $A \cap B$.

The **union** of two sets is the set that contains all the elements of both sets. The union of A and B is written $A \cup B$.

14. If $A = \{2, 3, 5, 7, 8, 9, 10\}$ and $B = \{3, 5, 8, 9\}$, is $A \subseteq B$? Is $B \subseteq A$?

_____

15. For the sets $A = \{x \mid x \leq 9, x \text{ is a natural number}\}$, $B = \{4, 6, 8\}$, $C = \{3, 5, 8, 10\}$,

a. Which of the sets A, B, C are subsets of set A?

b. Which pairs are disjoint?

c. Are any of these three sets equal?

_____

16. Is the set of integers a subset of the set of rational numbers?

_____

17. Are sets of rational numbers and sets of irrational numbers disjoint?

_____

In Exercises 18-22, consider the sets
$A = \{-4, -2, 0, 2, 6, 8\}$, $B = \{2, 3, 4, 5, 6\}$, $C = \{-3, -5, -7, 1, 5\}$.

18. Find $A \cap B$.

_____

19. Find $A \cup B$.

_____

20. Find $A \cap C$.

_____

21. Find $B \cap C$.

_____

Objective 5. Find absolute values

Main Idea

The distance a number a is from 0 on the number line (without regard to direction) is the **absolute value** of a, denoted $|a|$. The absolute value if any nonzero number is positive, and the absolute value of 0 is 0.

22. Compute: $|-4|$.

_____

23. Compute: $|5|$

_____

24. What is the distance from -9 to 0 on the number line?

_____

25. Compute: $|7-10|$

_____

Objective 6. Calculate with real numbers

Main Idea

Order of Operations:

1. Perform all operations inside parentheses or other grouping symbols before removing them.

2. Raise numbers to indicated powers and take indicated roots.

3. Do all multiplications and divisions, in the order they occur from left to right.

4. Do all additions and subtractions, in the order they occur from left to right.

To *add* two signed numbers with the *same sign*, add their absolute values (the numerical values, disregarding the sign), and keep their common sign.
Example. $|a|$ $(-4) + (-5) = -9$

To *add* two signed numbers with *unlike signs*, subtract the smaller absolute value from the larger absolute value, and keep the sign of the number with the larger absolute value.
Example. $(-7) + (3) = -(7 - 3) = -4$

To *subtract* two signed numbers, change the sign of the number being subtracted and proceed as in addition.   **Example.** $(-9) - (-3) = (-9) + 3 = -6$

To *add three or more signed numbers*, add them two at a time. If no grouping symbols are present, add from left to right. If grouping symbols are present, add inside the grouping symbols first.
Example. $-8 + (-6) - 4 - 3 = (-8 + (-6)) - 4 - 3 = -14 - 4 - 3 = -18 - 3 = -21$

When *multiplying or dividing* two numbers with the *same* sign, the product or quotient is positive.   **Example.** $(-2)(-3) = 6;$ $12 \div 3 = 4$

When *multiplying or dividing* two numbers with *unlike* signs, the product or quotient is negative.   **Example.** $(-5)(3) = -15;$ $\dfrac{16}{-2} = -8$

When *multiplying or dividing more than two signed numbers*, the product will be positive if there is an even number of negative signs. The product will be negative if there is an odd number of negative signs.

A number **multiplied by 0** is equal to 0.

Division by 0 is undefined. Zero divided by a non-zero number is defined.

26. Compute: $(-6)+(-16)$

_____

27. Compute: $19+(-18)$

_____

28. Compute: $22-(-13)$

_____

29. Compute: $-56-(-34)$

_____

30. Compute: $(-12)+(-6)-(-3)-6+(-14)$

_____

31. Compute: $(-7)(-3)(2)$

_____

32. Compute: $18\div(-2)+(-16)(-3)$

_____

33. Compute: $\dfrac{35}{-7}+\dfrac{-42}{-6}$

---

34. Compute: $\left[(-8)(-3)+5(-2)\right]\div\left[3(-4)-(-5)\right]$

---

35. Perform the following operations, if possible. If not possible, state why.

a. $(-3)(0)$ b. $\dfrac{0}{-6}$ c. $\dfrac{5}{0}$

Objective 7. Use properties of real numbers

Main Idea

1. The *Commutative Property of Addition* states that the sum of two real numbers is the same even if the order of the numbers is changed. That is, addition of real numbers is commutative. $a + b = b + a$ for all real numbers.

2. The *Commutative Property of Multiplication* states that the product of two real numbers is the same even if the order of the numbers is changed. That is, multiplication of real numbers is commutative. $a \bullet b = b \bullet a$ for all real numbers.

3. The *Associative Property of Addition* states that the sum of three numbers is the same if the first pair or the last pair are added first. That is, $a+(b+c)=(a+b)+c$.

4. The *Associative Property of Multiplication* states that the product of three numbers is the same if the first pair or the last pair are multiplied first. That is, $a\bullet(b\bullet c)=(a\bullet b)\bullet c$.

5. The *Distributive Property of Multiplication over Addition* states that multiplying a sum of two numbers by a third number gives the same result as multiplying the third number by each of the two numbers in the sum and adding the two products. That is, $a\bullet(b+c)=a\bullet b+a\bullet c$.

6. 0 is the *additive identity*. For any real number a, $a + 0 = a$.

7. 1 is the *multiplicative identity*. For any real number a, $a \cdot 1 = a$

8. For every real number a there exists a number b such that $a + b = 0$. The *additive inverse* of a is $-a$.

9. For every real number $a \neq 0$, there exists a number b such that $a \cdot b = 1$. The *multiplicative inverse* of a is also called the *reciprocal* of a and denoted $\dfrac{1}{a}$.

36. Use a commutative property to complete each statement.

a. $-8 + 6 = 6 +$ _____

b. $(-9)(3) = (3)$_____

_____

37. Use an associative property to complete each statement.

a. $8 + (-2) + 3 = 8 +$ _____

b. $((-7)(3)](-4) =$ _____(-4)

_____

38. Use the distributive property to rewrite each expression.

a. $6(12 + 5)$

b. $-(x + y)$

_____

39. Is the following statement an example of a commutative property or an associative property, or both?

$(3 + 7) + 8 = 7 + (3 + 8)$

_____

40. Use an identity property to complete each statement.

a. $-10 +$ _____ $= -10$

b. _____ $\cdot (3) = 3$

_____

41. Complete the statement.

a. $-4 + \underline{\hspace{1cm}} = 0$

b. $7 \bullet \underline{\hspace{1cm}} = 1$

Objective 8. Express inequalities as intervals and graph inequalities

Main Idea

An inequality is a statement that one quantity is greater (or less) than another quantity.

$a < b$ means that a lies to the left of b on the number line.

$a > b$ means that a lies to the right of b on the number line.

$a \leq b$ means a is less than or equal to b, and $a \geq b$ means that a is greater than or equal to b.

The subset of real numbers that lie between a and b (excluding a and b) can be written as the double inequality $a < x < b$ or by the open interval (a, b).

The closed interval [a, b] represents all numbers between a and b, including a and b, and can also be written $a \leq x \leq b$.

Intervals containing one endpoint, such as [a, b) are called half-open intervals.

We can represent the inequality $x > a$ by the interval (a, ∞) and the inequality $x < a$ by the interval $(-\infty, a)$.

Here are some examples of the graphs of different types of intervals.

Interval Notation	Inequality Notation	Number Line Graph
(a, ∞)	$x > a$	
$(-\infty, b)$	$x < b$	
(a, b)	$a < x < b$	

42. Express the graph as an inequality:

$$-8\,-7\,-6\,-5\,-4\,-3\,-2\,-1\ 0\ 1\ 2\ 3\ 4\ 5\ 6\ 7\ 8$$

_____

43. Express the interval as an inequality: $[-4, 3]$

_____

44. Express the inequality as an interval: $x \le -8$

_____

45. Express the inequality as an interval: $2 \le x \le 10$

_____

46. Express the graph as an interval:

$$-8\,-7\,-6\,-5\,-4\,-3\,-2\,-1\ 0\ 1\ 2\ 3\ 4\ 5\ 6\ 7\ 8$$

_____

47. Graph the inequality on a number line: $x \ge -2$

_____

48. Graph the inequality on a number line: $5 > x \ge -2$

_____

49. Graph the interval on a number line: $(-2, \infty)$

_____

50. Graph the interval on a number line: $[-5, 6)$

_____

Objective 9. Identify the coefficients of terms and constants in algebraic expressions

Main Idea

Letters representing unknown quantities are called **variables**, and fixed numbers (real numbers) are called **constants**. An expression created by performing additions, subtractions, or other arithmetic operations with one or more real numbers and variables is called an **algebraic expression**. A **term** of an algebraic expression is the product of one or more variables and a real number; the real number is called a **numerical coefficient** or simply a **coefficient**. A constant is also considered a term of an algebraic expression and is called a **constant term**. Examples of algebraic expressions are

$$8x - 7y, \ \frac{3z-5}{10+6y}, \ 6x^2 - 7x.$$

51. Identify the coefficient of each term and the constant term in the algebraic expression $-3x^2 - 4x + 8$.

_____

52. Identify the coefficient of each term and the constant term in the algebraic expression $5x^4 - 7x^3 - 3$

_____

Objective 10. Combine like terms

Main Idea Terms that contain exactly the same variables with exactly the same exponents are called **like terms**. For example, $3x^2y$ and $-7x^2y$ are like terms, but $3x^2y$ and $3xy$ are not like terms. To combine like terms, we combine (add or subtract) their coefficients. For example, $3x^2y+(-7x^2y)=(3+(-7))x^2y=-4x^2y$

53. Determine if the terms are like: $\dfrac{1}{2}xy^2$ and $-9xy^2$

_____

54. Determine if the terms are like: $-2x^3y$ and $5xy^3$

_____

55. Simplify the expression by combining like terms: $3x-8y+6x-4y$

_____

56. Simplify the expression by combining like terms: $2x^2-3xy-4y^2-5x^2+6xy-6y^2$

_____

Objective 11. Removing parentheses

Main Idea

We often need to remove parentheses when simplifying algebraic expressions and when solving equations. Removing parentheses often requires the use of the distributive property (which says that $a(b+c)=ab+ac$ for real numbers a, b, and c.) Care must be taken to avoid mistakes with signs when using the distributive property. Multiplying a sum in parentheses by a negative number changes the sign of each term in the parentheses. For example, $-2(x+4y)=(-2)(x)+(-2)(4y)=-2x-8y$.

57. Remove the parentheses and simplify: $4(p+d)$

_____

58. Remove the parentheses and simplify: $-2(3x-7y)$

_____

59. Remove the parentheses and simplify: $-a(b+8c)$

_____

60. Remove the parentheses and simplify: $4(x-3y)-(3x+2y)$

_____

61. Remove the parentheses and simplify: $4(2x-y)+4xy-5(y-xy)-(2x-4y)$

_____

Objective 12. Evaluate algebraic expressions

Main Idea Constants can be substituted for variables in order to evaluate expressions.

62. Find the value of the expression $2a - 4b$ when $a = 4$ and $b = -5$.

_____

63. Find the value of the expression $-2y + \dfrac{3}{x} - 5$ when $x = 6$ and $y = -7$.

_____

64. Find the value of the expression when $a = -2$, $b = 4$, and $c = 0.8$:

$3(2a - b) + 0.6(b - 3c)$

_____

65. Find the value of the expression if $x = 9$, $y = -7$, and $z = -3$:

$\dfrac{x^2 - 2y}{z(y - 3z)}$

_____

Objective 13. Plot points on a coordinate system

Main Idea To graph in two dimensions, we use a **rectangular (or Cartesian) coordinate system**. We construct the coordinate system by drawing a horizontal number line and a vertical number line so that they intersect at their origins. The point of intersection is called the **origin** of the system, the number lines are called the **coordinate axes**, and the plane is divided into four parts called **quadrants**. We call the horizontal axis the *x*-axis and the vertical axis the *y*-axis, and we denote any point in the plane as the ordered pair (x, y).

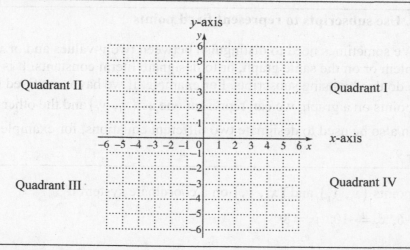

The ordered pair (a, b) represents the point P that is $|a|$ units from the *y*-axis (right if *a* is positive, and left if *a* is negative) and $|b|$ units from the *x*-axis (up if *b* is positive, and down if *b* is negative). The values of *a* and *b* are called the **rectangular coordinates** of the point. Note that if the point P is in the 1ˢᵗ quadrant, then *a* and *b* are both positive; if P is in the 2ⁿᵈ quadrant, then *a* is negative and *b* is positive; if P is in the third quadrant, then *a* and *b* are both negative; and if P is in the 4ᵗʰ quadrant, then *a* is positive and *b* is negative.

66. Plot the points on a rectangular coordinate system, and name the quadrant in which they lie.

A(−1, 3); B(5, −6); C(0, −8)

67. Name a point that lies in Quadrant II; in Quadrant III.

_____

Objective 14. Use subscripts to represent fixed points

Main Idea We sometimes need to distinguish between two y-values and/or x-values in the same problem or on the same graph, or to designate literal constants. It is often convenient to do this by using subscripts. For example, if we have two fixed but unidentified points on a graph, we can represent one as (x_1, y_1) and the other as (x_2, y_2). Subscripts can also be used to designate two different equations, for example $y_1 = 4x - 5$ and $y_2 = 6x + 9$.

---

68. Plot the points (x_1, y_1) and (x_2, y_2) on a coordinate system if $x_1 = 2$, $y_1 = -6$, $x_2 = -10$, $y_2 = 8$.

_____

Objective 1. Rounding and significant digits

Main Idea

Often we need to round a function to a certain number of decimal places. Recall the following:

Tenths means 0.1; hundredths means 0.01, thousandths means 0.001, ten-thousandths means 0.0001, and so on.

In some cases, however, we will be asked to report a function with three or four digits of accuracy, called **significant digits**. In general, significant digits are those from the first non-zero digit on the left to the last digit *after* the decimal point, if there is a decimal point. If there is no decimal point, any zeros to the right of the non-zero digits are not significant. Thus, 0.0752, 67,300 and 40.0 all have three significant digits.

In Exercises 1-5, determine the number of significant digits in each of the following numbers.

1.   52,865.3

_____

2.   0.003389

_____

3.   4700

_____

4.   4.0092

_____

5. 58.610

___

6. Rewrite the function $f(x) = 0.007543x^2 - 632.5778x + 480.65$ with the coefficients (and constant) rounded to:

a. three decimal places

___

b. three significant digits

___

c. four significant digits

___

Objective 2. Write fractions in lowest terms

Main Idea

A fraction is in lowest terms if the numerator and denominator have no common factors other than 1.

Example 1. The numerator and denominator of $\dfrac{25}{60}$ have a common factor of 5, so the fraction can be reduced by dividing both numerator and denominator by 5.

$$\frac{25}{60} = \frac{5 \cdot 5}{5 \cdot 12} = \frac{\cancel{5} \cdot 5}{\cancel{5} \cdot 12} = \frac{5}{12}$$

Example 2. You can simplify a fraction by writing the prime factorization of the numerator and denominator, and then dividing by the common factors. Recall that a prime number is a whole number greater than 1 whose only factors are 1 and itself.

$$\frac{60}{126} = \frac{2 \cdot 2 \cdot 3 \cdot 5}{2 \cdot 3 \cdot 3 \cdot 7} = \frac{\cancel{2} \cdot 2 \cdot \cancel{3} \cdot 5}{\cancel{2} \cdot 3 \cdot \cancel{3} \cdot 7} = \frac{10}{21}$$

In Exercises 7-12, write the fraction in lowest terms.

7. $\dfrac{21}{27}$

8. $\dfrac{20}{24}$

9. $\dfrac{77}{105}$

10. $-\dfrac{72}{112}$

11. $-\dfrac{120}{45}$

12. $\dfrac{150}{105}$

Objective 3. Multiply and divide fractions

Main Idea

To multiply two fractions, multiply their numerators and multiply their denominators. Often it is simpler to factor the numerator and denominator first, and then divide by common factors.

To divide fractions, multiply by the reciprocal of the divisor. Recall a quotient is the dividend divided by the divisor.

Example 1.

$$\frac{6}{7} \cdot \frac{21}{10} = \frac{6 \cdot 21}{7 \cdot 10} = \frac{2 \cdot 3 \cdot 3 \cdot 7}{7 \cdot 2 \cdot 5} = \frac{\cancel{2} \cdot 3 \cdot 3 \cdot \cancel{7}}{\cancel{7} \cdot \cancel{2} \cdot 5} = \frac{9}{5}$$

Example 2.

$$\left(-\frac{55}{18}\right) \div \frac{25}{4} = \left(-\frac{55}{18}\right) \cdot \frac{4}{25} = \frac{-5 \cdot 11 \cdot 2 \cdot 2}{2 \cdot 3 \cdot 3 \cdot 5 \cdot 5} = -\frac{\cancel{5} \cdot 11 \cdot \cancel{2} \cdot 2}{\cancel{2} \cdot 3 \cdot 3 \cdot \cancel{5} \cdot 5} = -\frac{22}{45}$$

In Exercises 13-18, multiply or divide the fractions, as indicated, and reduce to lowest terms.

13. $\dfrac{7}{5} \cdot \dfrac{10}{21}$

_____

14. $\dfrac{15}{22} \cdot \dfrac{11}{9}$

_____

15. $\dfrac{12}{13} \cdot \left(-\dfrac{65}{9}\right)$

_____

16. $\dfrac{11}{7} \div \dfrac{15}{7}$

_____

17. $\dfrac{9}{16} \div \dfrac{63}{20}$

_____

18. $\left(-\dfrac{405}{45}\right) \div \left(-\dfrac{90}{63}\right)$

Objective 4. Add and subtract fractions

Main Idea

To add or subtract fractions with *like* denominators, add or subtract the numerators, over their common denominator.

To add or subtract fractions with *unlike* denominators, first find the least common denominator of the fractions. Then rewrite each fraction so that it has the common denominator, using the Multiplication Property of One. Finally, add or subtract their numerators, over their common denominators.

Recall that to find the *least common denominator*, factor all the denominators. The LCD is the product of each different factor, used the maximum number of times it occurs in any one denominator. Equivalent fractions are found by multiplying the numerator and denominator by factors needed to obtain the LCD.

Example 1.

$$\dfrac{72}{25} - \dfrac{64}{25} = \dfrac{72-64}{25} = \dfrac{8}{25}$$

Example 2.

$$\dfrac{2}{3} + \dfrac{7}{10} - \dfrac{15}{6} = \dfrac{2}{3}\cdot\dfrac{10}{10} + \dfrac{7}{10}\cdot\dfrac{3}{3} - \dfrac{15}{6}\cdot\dfrac{5}{5} = \dfrac{20}{30} + \dfrac{21}{30} - \dfrac{75}{30} = \dfrac{20+21-75}{30} = -\dfrac{34}{30}$$

Note: The denominators are factored as $3,\ 2\cdot5,\ 2\cdot3$. The LCD is $2\cdot3\cdot5 = 30$

In Exercises 19-26, add or subtract the fractions, as indicated, and simplify the result.

19. $\dfrac{5}{12} + \dfrac{11}{12}$

20. $-\dfrac{17}{5}+\dfrac{13}{5}$

---

21. $\dfrac{7}{8}-\dfrac{15}{8}$

---

22. $\dfrac{9}{20}+\dfrac{4}{15}$

---

23. $\dfrac{8}{21}-\dfrac{9}{28}$

---

24. $-\dfrac{43}{33}+\dfrac{17}{22}$

---

25. $-\dfrac{13}{12}+\dfrac{11}{6}-\dfrac{49}{90}$

---

26. $\left(\dfrac{5}{7}-\dfrac{8}{21}\right)\div\left(-\dfrac{33}{6}+\dfrac{5}{14}\right)$

Objective 5. Identify expressions and equations

Main Idea

An algebraic expression contains constants and/or variables combined with additions, subtractions, multiplications, divisions, powers, or roots. An equation contains two algebraic expressions separated by the "=" symbol.

Recall some of the key words that indicate an operation:

Addition: added to, more than, the sum of, increased by, plus, the total of

Subtraction: subtracted from, minus, less, less than, the difference between, decreased by

Multiplication: times, the product of, multiplied by, of, twice, half

Division: divided by, the quotient, the ratio of

In Exercises 27-36, state identify each as an expression or an equation, then translate it to algebraic symbols.

27. 5 added to a number

_____

28. A number increased by 6, then multiplied by 2

_____

29. 15 less than a number

_____

30. 76 decreased by y

_____

31. The sum of m and p, divided by 6

_____

32.   Twice the sum of a number and 8 is equal to 64

_____

33.   The sum of two numbers is 50

_____

34.   The quotient of a number and 10 is 5 more than another number

_____

35.   The product of -7 and a number is 8 more than the number

_____

36.   Half the difference of a number and 8, decreased by 12

_____

Objective 6. Solve basic linear equations using the addition or subtraction properties

Main Idea

We can frequently solve an equation for a given variable by rewriting the equation in an equivalent form whose solution is easy to find. Two equations are equivalent if and only if they have the same solution.

Addition Property of Equations: Adding the same number to both sides of an equation gives an equivalent equation.

Subtraction Property of Equations: Subtracting the same number to both sides of an equation gives an equivalent equation.

Example 1. The equation $x - 8 = 7$ is equivalent to $x - 8 + 8 = 7 + 8$, or $x = 15$.

Example 2. The equation $p + 9 = -2$ is equivalent to $p + 9 - 9 = -2 - 9$, or $p = -11$.

In Exercises 37-40, state the property (or properties) of equations that can be used to solve each of the following equations; then use the property (or properties) to solve the equation.

37. $x + 7 = 11$

_____

38. $x + 5 = -2$

_____

39. $x - 4 = 10$

_____

40. $x - 9 = -5$

_____

Objective 7. Use the multiplication or division properties of equations to solve linear equations

Main Idea

Multiplication Property of Equations: Multiplying both sides of an equation by the same nonzero number gives an equivalent equation.

Division Property of Equations: Dividing both sides of an equation by the same nonzero number gives an equivalent equation.

Substitution Property of Equations: The equation formed by substituting one expression for an equal expression is equivalent to the original equation.

Example 1. The equation $\dfrac{y}{4} = -12$ is equivalent to $4\left(\dfrac{y}{4}\right) = 4(-12)$, or $y = -48$.

Example 2. The equation $-3x = 48$ is equivalent to $\dfrac{-3x}{-3} = \dfrac{48}{-3}$, or $y = -16$.

In Exercises 41-46, state the property (or properties) of equations that can be used to solve each of the following equations; then use the property (or properties) to solve the equation.

41.　　$3x = 6$

42.　　$5x = -20$

43.　　$-2x = 18$

44.　　$-4x = -24$

45.　　$\dfrac{y}{6} = 3$

46.　　$\dfrac{-p}{4} = -8$

Objective 8. Solve more linear equations

Main Idea

1. If the equation contains fractions, multiply both sides of the equation by the least common denominator of the fractions.

2. Remove any parentheses or other grouping symbols.

3. Perform any additions or subtractions to get all terms containing the variable on one side and all other terms on the other side of the equation. Combine like terms.

4. Divide both sides by the coefficient of the variable.

Solve the equations in Exercises 47-56.

47. $2x + 8 = -12$

48. $\dfrac{x}{4} - 3 = 6$

49. $4x - 3 = 6 + x$

50. $3x - 2 = 4 - 7x$

51. $\dfrac{3x}{4} = 12$

---

52. $\dfrac{5x}{2} = -10$

---

53. $3(x-5) = -2x-5$

---

54. $-2(3x-1) = 4x-8$

---

55. $2x-7 = -4\left(4x-\dfrac{1}{2}\right)$

---

56. $-2(2x - 6) = 3\left(3x - \dfrac{1}{3}\right)$

In Exercises 57-60, use the Substitution Property of Equations to solve the equation.

57. Solve for x if $y = 2x$ and $x + y = 12$

58. Solve for x if $y = 4x$ and $x - y = 24$

59. Solve for x if $y = -3x$ and $2x + 4y = 20$

60. Solve for x if $y = -6x$ and $2x - 2y = -20$

Objective 9. Determine if a linear equation is a conditional equation, an identity, or a contradiction

Main Idea

An equation is a statement that two quantities or expressions are equal. An equation is an *identity* if it is true for all values of the variable for which both sides of the equation are defined. An equation that is true for some values of the variables but is not an identity is called a *conditional equation*. A conditional equation is usually called an equation. An equation that is not true for any values of the variables is called a *contradiction*.

Example 1. The equation $3x = 12$ is true when $x = 4$, so it is a conditional equation.

Example 2. The equation $7x + 4x = 13x - 2x$ is true for all values of x, so it is an identity.

Example 3. The equation $2(x - 3) = 2x + 1$ is not true for any value of the variable, so it is a contradiction.

In Exercises 61-64, determine whether the equation is a conditional equation, an identity, or a contradiction.

61. $3x - 5x = 2x + 7$

_____

62. $3(x + 1) = 3x - 7$

_____

63. $9x - 2(x - 5) = 3x + 10 + 4x$

_____

64. $\dfrac{x}{2} - 5 = \dfrac{x}{4} + 2$

_____

Objective 10. Evaluate formulas for given values

Main Idea   A formula contains one or more letters that represent constants or variables.

In Exercises 65-71, solve each formula as requested.

65.   Given the formula $A = lw$, if $l = 7$ and $w = 6$, find A.

66.   Solve for P in the formula $A = Prt$ if $A = 3000$, $r = 0.06$, and $t = 5$.

67.   Solve for h in the formula $A = \frac{1}{2}bh$ if $A = 320$ and $b = 40$.

68.   If $V = \frac{1}{3}Bh$, solve for B if $V = 900$ and $h = 45$.

69.   If $C = 2\pi r$, solve for C if $r = 9$.

_____

70.   Solve for t in the formula $S = P + Prt$, if $P = 5000$, $r = 0.05$, and $S = 7000$

_____

71.   Solve for C in the formula $5F - 9C = 160$, if $F = 75$.

_____

Objective 11. Use properties of inequalities to solve linear inequalities

Main Idea

As with equations, we can find solutions to inequalities by finding equivalent inequalities from which the solutions can be easily seen. The following properties will give an equivalent inequality:

Substitution Property: Substituting one expression for an equal expression

Addition and Subtraction Property: Adding the same quantity to (or subtracting the same quantity from) both sides of the inequality

Multiplication Property I: Multiplying (or dividing) both sides of an inequality by the same *positive* number

Multiplication Property II: Multiplying (or dividing) both sides of an inequality by the same *negative* number and reversing the inequality symbol

---

In Exercises 72-79, use Properties of Inequalities to solve the inequalities.

72. $5x + 1 > -5$

_____

73. $1 - 3x \geq 7$

_____

74. $\dfrac{x}{4} > -3$

_____

75. $\dfrac{x}{6} < -2$

_____

76. $\dfrac{x}{3} - 2 > 5x$

_____

77. $\dfrac{x}{2} + 3 \le 6x$

_____

78. $-3(x-5) < -4$

_____

79. $-\dfrac{1}{2}(x+4) < 6$

_____

Objective 1. Use the properties of exponents with integers

Main Idea

Product Rule of Exponents: $a^m \cdot a^n = a^{m+n}$
The product of two exponential numbers with the same base is the base raised to the sum of the exponents.

Power Rule of Exponents: $\left(a^m\right)^n = a^{mn}$

An exponential number raised to a power is the base raised to the product of the exponents.

Power Rule for Products: $\left(ab\right)^m = a^m b^m$

A product of two bases raised to a power is the product of each base raised to the power.

Power Rule for Quotients: $\left(\dfrac{a}{b}\right)^m = \dfrac{a^m}{b^m}$

A quotient of two bases raised to a power is the quotient of each base raised to the power.

Zero Exponent: For any nonzero real number a, we define $a^0 = 1$. We leave 0^0 undefined.

Negative Exponents: $a^{-1} = \dfrac{1}{a}$ and $a^{-n} = \dfrac{1}{a^n}$ for nonzero a.

Use rules of exponents to simplify the following expressions. Leave answers with positive exponents.

1. $x^3 \cdot x^5$

2. $\left(x^{-4}\right)\left(x^6\right)$

---

3. $6^3 \cdot 6^{-2}$

4. $a^5 \cdot a$

---

5. $y^{-5} \cdot y^2$

6. $2^{-3} \cdot 2^{-4}$

7. $\dfrac{x^8}{x^4}$

8. $\dfrac{a^5}{a^{-1}}$

9. $\dfrac{y^{-3}}{y^{-4}}$

10. $\left(x^4\right)^3$

11. $(xy)^2$

12. $\left(x^2\right)^{-3}$

13. $(2m)^3$

14. $\left(2x^{-2}y\right)^{-4}$

15. $\left(\dfrac{x^2}{y^3}\right)^5$

16. $\left(-32x^5\right)^{-2}$

---

17. 16^0

18. $\left(\dfrac{x^{-2}y}{z}\right)^{-3}$

---

19. $\left(\dfrac{a^{-2}b^{-1}c^{-4}}{a^4b^{-3}c^0}\right)^{-3}$

Objective 2. Understand terminology related to polynomials

Main Idea

An algebraic expression containing a finite number of additions, subtractions, and multiplications of constants and nonnegative integer powers of variables is called a **polynomial**. When simplified, a polynomial cannot contain negative powers of variables, fractional powers of variables, variables in a denominator, or variables inside a radical. Examples of polynomials are:

$$5x - 2y, \quad 7z^3 + 2x, \quad \frac{1}{3}x^4 - 8x^2 - 3x$$

Examples that are not polynomials are:

$$\frac{3x - 5}{13 + 5y}, \quad 3x^2 - 6\sqrt{x}, \quad 5a^{-5} + 7a^2 - 3$$

A **monomial** is a polynomial with exactly one term. A **binomial** is a polynomial with exactly two terms. A **trinomial** is a polynomial with exactly three terms. The **degree of a term** is the sum of the exponents of every variable in that term. The **degree of a polynomial** is the degree of the highest degree term in the polynomial. The degree of a constant is 0. A polynomial is written in **descending order** if the powers of one of its variables decrease from left to right. The **leading coefficient** of a polynomial is the coefficient of the term with highest degree.

Exercises

20. Decide if each of the following is a polynomial.

a. $3x^4 - 7x^2 + 3$
b. $-4x^{-3} + 6x$
c. $\dfrac{s^2 - 5s}{3t + 6}$

_____

21. Label each of the following as a monomial, binomial, or trinomial.

a. $18z$
b. $-5y^2 + 4y - 7$
c. $16ab^3cx$
d. $10 - 4x^2$

_____

22. State the degree of each term.

a. $3x^2$
b. $6xy^3$
c. 16
d. $-8k$

_____

23. State the degree and leading coefficient of each polynomial.

a. $-5y^2 + 4y - 7$
b. $4x^2 - 6x^4 + 3x - 2$
c. $-x + 3$

_____

24. Write the polynomial in descending order.

$6x^2 - x^4 + 10 + 3x$

_____

Objective 3. Add and subtract monomials and polynomials

Main Idea   To add or subtract polynomials:

(a) If no parentheses are present, then combine like terms.

(b) If the parentheses are preceded by a plus sign, then remove the parentheses without changing signs. Then combine like terms.

(c) If the parentheses are preceded by a negative sign, then remove the parentheses and change the sign of every term that was inside the parentheses. Then combine like terms.

25. Simplify $7x + 3x - 8x$

_____

26. Simplify $3x^2 - 9x + 2 + 5x^2 - 6x - 8$

_____

27. Simplify $x + (3x + 2) - (2x - 3)$

_____

28. Simplify $\left(3y^3 - 6y + 1\right) + \left(7y^2 - 10y - 9\right)$

_____

29. Simplify $\left(5p^4 + 6p^2 + 5\right) - \left(7p^4 + 8p^2 - 8p - 3\right)$

_____

30. Subtract $\left(-2c^3 - 2c^2 + 2\right)$ from $\left(2c^3 + 7c^2 - 8c - 1\right)$ and simplify.

_____

31. Simplify $-\left[\left(6v^3+7v^2-7\right)-\left(4v+6v^2-2v^3\right)\right]+\left(3v^2-8v+2\right)$

Objective 4. Multiply monomials and polynomials

Main Idea

We can multiply monomials using the rules of exponents and the commutative and associative properties for multiplication. For example,
$$\left(2y^2z\right)\left(-3xyz^2\right)=2(-3)x\left(y^2y\right)\left(zz^2\right)=-6xy^3z^3.$$

We can use the distributive property to multiply a binomial by a monomial. For example, $x(3x+2)=x\cdot3x+x\cdot2=3x^2+2x$. We can use the extended distributive property to multiply a polynomial by a monomial. For example,
$$5\left(x+y^2+2z\right)=5\cdot x+5\cdot y^2+5\cdot2z=5x+5y^2+10z.$$

The distributive property can also be used to multiply a polynomial by a polynomial. In the case of multiplying two binomials, we have
$(a+b)(c+d)=a(c+d)+b(c+d)=ac+ad+bc+bd$, which is the sum of the products of the First, Outside, Inside, and Last terms of the binomials, known as FOIL. In general, when multiplying two binomials, multiply each term of one polynomial by every term of the other polynomial.

Multiply and simplify the following:

32. $\left(5x^3\right)\left(7x^2\right)$

---

33. $\left(-3x^2y\right)\left(2xy^3\right)\left(4x^2y^2\right)$

---

34. $(3mx)(2mx^2)-(4m^2x)x^2$

_____

35. $ax^2(2x^2+ax-ab)$

_____

36. $(4a+5b-6c)ac$

_____

37. $(x-4)(x+3)$

_____

38. $(3x+2)(2x-5)$

_____

39. $\left(1-2x^2\right)\left(2-x^2\right)$

_____

40. $(a-2b)\left(a^2-3ab+b^2\right)$

_____

Objective 5. Divide polynomials

Main Idea

To divide a polynomial by a monomial, divide the monomial into each term of the polynomial. Use the rules of exponents to simplify each of the quotients. If the minimal is not a factor of one of the terms, reduce the resulting fraction to lowest terms.

Example. $\dfrac{6x^5-4x^4+2}{2x}=\dfrac{6x^5}{2x}-\dfrac{4x^4}{2x}+\dfrac{2}{2x}=3x^4-2x^3+\dfrac{1}{x}$

To divide one polynomial into another (long division):

1. Write the division problem with both polynomials in descending powers of one variable.

2. Divide the highest power of the divisor into the highest power of the dividend to obtain the first partial quotient.

3. Write this partial quotient above the highest power in the dividend. Multiply the divisor by this quotient; write the product under the dividend; and subtract like terms.

Example.
Divide:
$\left(4x^2+3x+2\right)\div(x+2)$

1.
$$x+2\overline{)4x^2+3x+2}$$

2. x divided into $4x^2$ is $4x$

3.
$$\begin{array}{r} 4x \\ x+2\overline{)4x^2+3x+2} \\ \underline{4x^2+8x} \\ -5x \end{array}$$

4. To the remainder bring down the next term of the dividend to form a new partial dividend. Divide the highest power of the divisor into the highest power of
the dividend and write this partial quotient above the dividend. Multiply the partial quotient times the divisor; write the product under the partial dividend; and subtract.

4.
$$\begin{array}{r} 4x-5 \\ x+2\overline{)4x^2+3x+2} \\ \underline{4x^2+8x} \\ -5x+2 \\ \underline{-5x-10} \\ 12 \end{array}$$

5. Repeat step 4 until all terms of the dividend have been used. Any remainder is written over the divisor.

5. All terms have been used. The quotient is
$$4x-5+\frac{12}{x+2}$$

---

41. Divide: $\dfrac{x^4+6x^3+5x^2-4x+2}{x+2}$

---

42. Divide: $\left(x^5+x^3-1\right)\div(x+1)$

---

43. Divide $4x^3+4x^2+5$ by $2x^2+1$

---

44. Divide: $\left(3x^5 - x^4 + 5x - 1\right) \div \left(x^2 - 2\right)$

Objective 6. Factor out common factors

Main Idea

Factoring is the process of writing a number or algebraic expression as the product of two or more different numbers or expressions. If a polynomial has integer coefficients, we can use the distributive property to factor out any common integer factors. For example, $2x^2 + 6 = 2\left(x^2 + 3\right)$. We can also use the distributive property to factor out any algebraic expression that is a factor of each term of an algebraic expression. Such a factor is called a **common factor**. And our goal is to remove the **largest possible** common factor. For example, to factor $2x^2y + 2xy + 8xy^2$, note that 2xy is a factor to all three terms, so $2x^2y + 2xy + 8xy^2 = 2xy(x + 1 + 2y)$.

Factor each of the following:

45. $9ab - 12ab^2 + 18b^2$

---

46. $8x^2y - 160x + 48x^2$

---

47. $x^3 - x^4$

_____

48. $12y^3z + 4yx^2 - 8y^2z^3$

_____

49. $x^2\left(x^2+4\right)^2 + 2\left(x^2+4\right)^3$

_____

Objective 7. Factor by grouping

Main Idea

Some polynomials, such as $5x - 5y + bx - by$, do not contain common factors, but are such that common factors can be factored from two (or more) terms. When these are factored, the polynomial is changed into a form that now contains a (new) common factor in each term. This is called **factoring by grouping**. For example, in $5x - 5y + bx - by$, 5 is a common factor of the first two terms and b is a common factor of the last two terms. So $(5x - 5y) + (bx - by) = 5(x - y) + b(x - y)$. The new expression has two terms, $5(x - y)$ and $b(x - y)$, both of which contain the factor $(x - y)$. Treating $(x - y)$ as a common factor gives $5(x - y) + b(x - y) = (x - y)(5 + b)$.

In each of the following, factor by grouping.

50. $3x - 3y + cx - cy$

_____

51. $9x^2 - 15x + 6x - 10$

_____

52. $x^3 + x^2 - x - 1$

_____

53. $x^3 - x^2 - 5x + 5$

_____

Objective 8. Factor polynomials

Main Idea

There are several useful special factorization formulas.

Difference of two squares: $a^2 - b^2 = (a+b)(a-b)$

Perfect square trinomial: $a^2 + 2ab + b^2 = (a+b)^2$

$$a^2 - 2ab + b^2 = (a-b)^2$$

Completely factor each of the following:

54. $x^2 - 16$

55. $25x^2 - y^2$

_____

56. $16z^2 - 81x^2$

57. $2x - 18x^3$

58. $x^2 + 8x + 16$

59. $x^2 - 4x + 4$

60. $4x^2 + 20x + 25$

61. $x^4 - 6x^2y + 9y^2$

62. $5x^5 - 80x$

Objective 9. Factor trinomials

Main Idea

To factor a trinomial into the product of its binomial factors:

1. If the leading coefficient is 1, and the trinomial is in descending order, then find two factors of the last term that add to be the middle coefficient. Then, factor by "reverse FOIL" using $(x+a)(x+b) = x^2 + (a+b)x + ab$. For example, $x^2 + 5x + 6 = (x+3)(x+2)$. Note that $3 \cdot 2 = 6$ and $3 + 2 = 5$.

2. If the leading coefficient is not 1, write the trinomial in descending order and factor out any greatest common factor, including -1, if possible. Determine two factors of the first term and begin to write binomial factors. Then use sign analysis: if all the signs of trinomial are positive, then signs between the terms of the binomial factors will be positive. If the sign of the third term is positive, but the sign of the middle term is negative, then the signs between the terms of the binomials will be negative. If the sign of the third term is negative, then the signs between the terms of the binomials will be opposite. Then find a pair of factors of the third term that, when the binomials are multiplied by FOIL, the sum of the outer and inner products add to be the middle term of the trinomial. Check by FOIL.

For example, to factor $3x^2 - 7x + 4$, first note that the trinomial is in descending order and there is no common factor.

We proceed by forming two binomial factors using factors of $3x^2$: $(3x \quad)(x \quad)$. Possible factors of 4 are 1 and 4 or 2 and 2. Since the second term is negative, and the third term is positive, the factors must both be negative. Choosing factors -1 and -4 in the arrangement $(3x-4)(x-1)$ works, because the outer product $-3x$ plus the inner product $-4x$ add to the middle term of the trinomial, $-7x$. Thus, the factorization of $3x^2 - 7x + 4$ is $(3x-4)(x-1)$.

Completely factor each of the following trinomials.

63. $x^2 - x - 6$

_____

64. $x^2 + 11x + 10$

_____

65. $2x^2 - x - 1$

___

66. $2x^2 - 8x - 42$

___

67. $3x^2 - x - 2$

___

68. $12x^2 + 11x + 2$

___

Objective 10. Radicals

Main Idea

The number b is a square root of a number a if $b^2 = a$. The principal square root of a nonnegative number is its nonnegative square root. The symbol $\sqrt{a}$ represents the principal square root of a. The expression written under the radical symbol is called the radicand, and the root is called the index. The square root of a negative number is not a real number. The number c is the cube root of a number a if $c^3 = a$. The symbol $\sqrt[3]{a}$ represents the cube root of a. The cube root of a positive number is a positive number, and the cube root of a negative number is a negative number. In general, the (principal) n^{th} root of a real number is defined as $\sqrt[n]{a} = b$ if $b^n = a$, and the even root of a negative number is not a real number. For example, $\sqrt[3]{64} = 4$ because $4^3 = 64$. Also $\sqrt[3]{-27} = -3$ because $(-3)^3 = -27$. $\sqrt{81} = 9$ because $9^2 = 81$. $\sqrt{-64}$ is not a real number because there is no real number that, when squared, gives -64. We note that $\sqrt[n]{a^n} = |a|$, and if a is nonnegative, $\sqrt[n]{a^n} = a$.

The Product Rule for Radicals states that for nonnegative real numbers a and b, $\sqrt{a} \cdot \sqrt{b} = \sqrt{ab}$ and $\sqrt{ab} = \sqrt{a} \cdot \sqrt{b}$, and in general, $\sqrt[n]{a} \cdot \sqrt[n]{b} = \sqrt[n]{ab}$ and $\sqrt[n]{ab} = \sqrt[n]{a} \cdot \sqrt[n]{b}$, provided $\sqrt[n]{a}$ and $\sqrt[n]{b}$ are real. We can use this property to simplify some radicals. For example, $\sqrt{20} = \sqrt{4 \cdot 5} = \sqrt{4} \cdot \sqrt{5} = 2\sqrt{5}$, and $\sqrt[3]{54} = \sqrt[3]{27 \cdot 2} = \sqrt[3]{27} \cdot \sqrt[3]{2} = 3\sqrt[3]{2}$.

In problems 69-72, find the roots, if they are real.

69. $\sqrt{121}$

70. $\sqrt[6]{64}$

71. $\sqrt[3]{-8}$

72. $\sqrt{-16}$

Main Idea

A radical is in simplified form when all of the following are true.

1. All possible operations are performed.
2. The radicand contains no factor (other than 1) that is a perfect power of the index.
3. There are no fractions in the radicand.
4. There are no radicals in a denominator.

To simplify a radical expression, look for factors of the radicand that are perfect squares if the radical is $\sqrt{\ }$, perfect cubes if the radical is $\sqrt[3]{\ }$, and so on. Rewrite the radicand using these perfect powers. Write the root of the perfect powers outside the radical.

In problems 73-76, simplify the radicals, assuming the expressions are real and the variables represent nonnegative real numbers.

73. $\sqrt{48x^3}$

_____

74. $\sqrt[3]{72a^3b^4}$

_____

75. $\sqrt{8t^2s^5}$

_____

76. $\sqrt[3]{-128a^4b^5c^6}$

Objective 12. Multiply and divide radicals

Main Idea

Recall that the Product Rule for Radicals states that for nonnegative real numbers a and b, $\sqrt{a} \cdot \sqrt{b} = \sqrt{ab}$, and in general, $\sqrt[n]{a} \cdot \sqrt[n]{b} = \sqrt[n]{ab}$, provided $\sqrt[n]{a}$ and $\sqrt[n]{b}$ are real. This allows us to multiply radicals with the same index. The Quotient Rule for Radicals states that $\dfrac{\sqrt[n]{a}}{\sqrt[n]{b}} = \sqrt[n]{\dfrac{a}{b}}$, provided $\sqrt[n]{a}$ and $\sqrt[n]{b}$ are real and $b \neq 0$

In problems 77-82, multiply or divide, as indicated, and simplify the result, assuming the expressions are real and the variables represent nonnegative real numbers.

77. $\sqrt{3} \cdot \sqrt{27}$ 78. $\sqrt{2a} \cdot \sqrt{6a}$

---

79. $\sqrt{8xy^3z} \cdot \sqrt{4x^2y^2z}$

---

80. $\sqrt[3]{2xy} \cdot \sqrt[3]{4x^2y}$

---

81. $\dfrac{\sqrt[3]{32}}{\sqrt[3]{4}}$

---

82. $\dfrac{\sqrt{16a^3x}}{\sqrt{2ax}}$

Objective 13.   Add and subtract radicals

Main Idea

We can add or subtract radicals if they are like radicals, that is, if they have the same index and the same radicand. We add (or subtract) like radicals by adding (or subtracting) their coefficients, as we add like terms in algebra. For example, $\sqrt{6} + 3\sqrt{6} = 4\sqrt{6}$. Before adding or subtracting radicals, be sure they are simplified.

83. Add: $2\sqrt{x} + 5\sqrt{x}$

84. Subtract: $5\sqrt[3]{ab} - 2\sqrt[3]{ab}$

---

85. Combine: $3\sqrt{8} + 2\sqrt{2}$

86. Combine: $3\sqrt{27} + 6\sqrt{3} - \sqrt{12}$

---

87. Combine: $4x\sqrt[3]{xy} + \sqrt[3]{8x^4y}$

Objective 14.   Convert between exponential and radical form

Main Idea

For a positive integer n, we define $a^{1/n} = \sqrt[n]{a}$ if $\sqrt[n]{a}$ exists .

Then, for any integer m, $a^{m/n} = \left(a^{1/n}\right)^m = \left(\sqrt[n]{a}\right)^m$.

And $a^{m/n} = \left(a^m\right)^{1/n} = \sqrt[n]{a^m}$ if a is nonnegative when n is even and if $a \neq 0$ when $m \leq 0$.

88. Simplify, if possible:

a. $4^{1/2}$

b. $(-27)^{1/3}$

c. $-16^{1/4}$

d. $(-4)^{1/2}$

---

89. Write the following in radical form and simplify:

a. $16^{3/4}$

b. $y^{-3/2}$

c. $(6m)^{2/3}$

---

90. Write the following without radicals:

a. $\sqrt{x^3}$

b. $\dfrac{1}{\sqrt[3]{b^2}}$

c. $\sqrt{(ab)^3}$

Objective 1. Write rational expressions in lowest terms

Main Idea

We can simplify rational expressions (algebraic fractions) by dividing the numerator and denominator by the same factor, using the property $\dfrac{ac}{bc} = \dfrac{a}{b}$, for $b \neq 0$, $c \neq 0$. Begin by completely factoring the numerator and denominator of the rational expression.

Example. $\dfrac{x^2 - 7x - 12}{x^2 - 9} = \dfrac{(x-3)(x-4)}{(x-3)(x+3)} = \dfrac{\cancel{(x-3)}(x-4)}{\cancel{(x-3)}(x+3)} = \dfrac{(x-4)}{(x+3)}$

Simplify:

1. $\dfrac{4x}{4x+8}$

2. $\dfrac{x-3y}{3x-9y}$

3. $\dfrac{x^2-6x+8}{x^2-16}$

4. $\dfrac{x^2 y^2 - 4x^3 y}{x^2 y - 2x^2 y^2}$

5. $\dfrac{3x^2 - 7x - 6}{x^2 - 4x + 3}$

Objective 2.   Multiply and divide rational expressions

Main Idea

To multiply two fractions, first factor the numerators and denominators. Then write the product of the fractions as a fraction with the numerator as an indicated product of the numerators and the denominator as an indicated product of the denominators. Simplify the fraction by dividing the numerator and denominator by their common factors.

Example.

$$\dfrac{x-3}{x^2-5x+4} \cdot \dfrac{x^2-3x+2}{x^2-9} = \dfrac{x-3}{(x-4)(x-1)} \cdot \dfrac{(x-2)(x-1)}{(x+3)(x-3)}$$

$$= \dfrac{(x-3)(x-2)(x-1)}{(x-4)(x-1)(x+3)(x-3)} = \dfrac{x-2}{(x-4)(x+3)}$$

To divide two fractions, invert the divisor and multiply as above.

Example.

$$\dfrac{x^2+7x+12}{x-2} \div \dfrac{x^2-9}{x^2-x-2} = \dfrac{x^2+7x+12}{x-2} \cdot \dfrac{x^2-x-2}{x^2-9} = \dfrac{(x+4)(x+3)}{x-2} \cdot \dfrac{(x-2)(x+1)}{(x+3)(x-3)}$$

$$= \dfrac{(x+4)(x+3)(x-2)(x+1)}{(x-2)(x+3)(x-3)} = \dfrac{(x+4)(x+1)}{x-3}$$

Multiply or divide, as indicated, and simplify the result:

6. $\dfrac{6x^3}{8y^3} \cdot \dfrac{16x}{9y^2} \cdot \dfrac{15y^4}{x^3}$

---

7. $\dfrac{x-2}{4x} \cdot \dfrac{12x^2}{x^2+x-6}$

---

8. $\dfrac{4x+4}{x-4} \cdot \dfrac{x^2-6x+8}{8x^2+8x}$

9. $\dfrac{16}{x-2} \div \dfrac{4}{3x-6}$

10. $\dfrac{y^2-2y+1}{7y^2-7y} \div \dfrac{y^2-4y-3}{35y^2}$

11. $\left(x^2-x-6\right) \div \dfrac{9-x^2}{x^2+3x}$

12. $\dfrac{2x^2+7x+3}{4x^2-1} \div \left(x+3\right)$

13. $\dfrac{x^2+x}{x^2-5x+6} \cdot \dfrac{x^2-2x-3}{2x+4} \div \dfrac{x^3-3x^2}{4-x^2}$

Objective 3.   Add and subtract rational expressions with the same denominator

Main Idea

To combine (add or subtract) fractions with the same denominator, write the sum or difference of the numerators over the common denominator. Simplify the numerator by removing parentheses and combining like terms. Check to see if any factors of the numerator are also factors of the denominator, and if so, reduce the fraction.

Combine the following fractions and simplify, if possible.

14. $\dfrac{6x-2}{3xy} + \dfrac{3x+2}{3xy}$

15. $\dfrac{2x+3}{x-9} + \dfrac{3x+7}{x-9}$

16. $\dfrac{2x+3}{x^2-1} + \dfrac{x+2}{x^2-1}$

Objective 4.   Find the least common denominator

Main Idea

To find the least common denominator of two or more fractions, completely factor each denominator, including finding the prime factors of the coefficients. The LCD is the product of each different factor of the denominators, each factor used the maximum number of times it occurs in any one denominator.

Example.

1. To find the LCD of $\dfrac{1}{5x}, \dfrac{2}{15x^2}, \dfrac{3}{20xy^2}$, factor the denominators: $5x$, $3 \cdot 5x^2$, $2^2 \cdot 5xy^2$.

The factor 2 is used at most twice, the 3 is used at most once, the 5 is used at most once, the x is used at most twice, and the y is used at most twice. Thus, the LCD is
$2^2 \cdot 3 \cdot 5 \cdot x^2 \cdot y^2 = 60x^2y^2$.

2. To find the LCD of $\dfrac{1}{x^2-x}, \dfrac{3}{x^2-1}, \dfrac{y}{3x^2}$, factor the denominators:

$x(x-1)$, $(x+1)(x-1)$, $3x^2$. The factor 3 is used at most once, the factor x is used at most twice, the factor $(x + 1)$ is used at most once, and the factor $(x - 1)$ is used at most once. Thus the LCD is $3x^2(x+1)(x-1)$.

Find the least common denominator for the fractions in the following problems.

17. $\dfrac{5x}{abc}, \dfrac{3y}{ab^2c^2}, \dfrac{7}{a^2b}$

18. $\dfrac{1}{x^2y}, \dfrac{3}{2xy^2}, \dfrac{-4}{x-1}$

Objective 5.   Add and subtract rational expressions with unlike denominators

Main Idea

To add or subtract rational expressions with unlike denominators, we must first express each fraction equivalently with the least common denominator. Then write the sum or difference of the numerators over the common denominator. Simplify the numerator by removing parentheses and combining like terms. Check to see if any factors of the numerator are also factors of the denominator, and if so, reduce the fraction.

Example.

$$\frac{x-3}{x-5}+\frac{2x-1}{x^2-x-20}=\frac{x-3}{x-5}-\frac{2x-1}{(x-5)(x+4)}=\frac{(x-3)(x+4)}{(x-5)(x+4)}-\frac{2x-1}{(x-5)(x+4)}$$

$$=\frac{(x-3)(x+4)-(2x-1)}{(x-5)(x+4)}=\frac{x^2+x-12-2x+1}{(x-5)(x+4)}=\frac{x^2-x-11}{(x-5)(x+4)}$$

Combine the following fractions and simplify, if possible.

19. $1+\dfrac{1}{x}-\dfrac{2}{x^2}$

20. $\dfrac{a}{a-2}-\dfrac{a-2}{a}$

21. $\dfrac{x-1}{x+1}-\dfrac{2}{x^2+x}$

22. $\dfrac{x-7}{x^2-9x+20}+\dfrac{x+2}{x^2-5x+4}$

23. $\dfrac{3x^2}{x^2-4}+\dfrac{2}{x^2-4x+4}-3$

Objective 6. Simplify complex fractions

Main Idea

To simplify a complex fraction, find the LCD of all fractions contained in the numerator and denominator of the complex fraction. Multiply both the numerator and denominator of the complex fraction by the LCD. (By the distributive property, this means that every term of the complex fraction will be multiplied by the LCD.) Simplify each term of the complex fraction either by dividing out common factors or by multiplying. Factor the resulting numerator and denominator and simplify, if possible.

Example.

$$\dfrac{\dfrac{1}{3}+\dfrac{4}{x}}{3-\dfrac{1}{xy}}=\left[\dfrac{\dfrac{1}{3}+\dfrac{4}{x}}{3-\dfrac{1}{xy}}\right]\cdot 3xy=\dfrac{\left[\dfrac{1}{3}+\dfrac{4}{x}\right]\cdot 3xy}{\left[3-\dfrac{1}{xy}\right]\cdot 3xy}=\dfrac{\dfrac{1}{3}\cdot 3xy+\dfrac{4}{x}\cdot 3xy}{3\cdot 3xy-\dfrac{1}{xy}\cdot 3xy}=\dfrac{xy+12y}{9xy-3}=\dfrac{y(x+12)}{3(3xy-1)}$$

Additional method: Rewrite the complex fraction with the main fraction bar written as a division symbol. If more than one fraction appears in the dividend (numerator of the complex fraction) or divisor (denominator of the complex fraction), find the LCD for each. Write the equivalent of each fraction with the desired LCD as its denominator. Add or subtract the fractions in the numerator and denominator so that the numerator contains a single fraction and the denominator contains a single fraction. Divide the fractions by inverting the divisor and multiplying. Write the indicated product of the numerators and the indicated product of the denominators. Simplify the fraction if possible.

Example.

$$\dfrac{\dfrac{1}{3}+\dfrac{4}{x}}{3-\dfrac{1}{xy}}=\left[\dfrac{1}{3}+\dfrac{4}{x}\right]\div\left[3-\dfrac{1}{xy}\right]=\left[\dfrac{1}{3}\cdot\dfrac{x}{x}+\dfrac{4}{x}\cdot\dfrac{3}{3}\right]\div\left[3\cdot\dfrac{xy}{xy}-\dfrac{1}{xy}\right]=\left[\dfrac{x}{3x}+\dfrac{12}{3x}\right]\div\left[\dfrac{3xy}{xy}-\dfrac{1}{xy}\right]$$

$$=\dfrac{x+12}{3x}\div\dfrac{3xy-1}{xy}=\dfrac{x+12}{3x}\cdot\dfrac{xy}{3xy-1}=\dfrac{(x+12)xy}{3x(3xy-1)}=\dfrac{y(x+12)}{3(3xy-1)}$$

In the following problems, simplify the complex fraction.

24. $$\dfrac{\dfrac{1}{x}+\dfrac{1}{y}}{\dfrac{1}{x}-\dfrac{1}{y}}$$

25. $$\dfrac{\dfrac{5}{2y}+\dfrac{3}{y}}{\dfrac{1}{4}+\dfrac{1}{3y}}$$

26. $$\dfrac{2-\dfrac{1}{x}}{2x-\dfrac{3x}{x+1}}$$

27. $\dfrac{1 - \dfrac{2}{x-2}}{x-6 + \dfrac{10}{x+1}}$

Objective 7. Work with the complex number i

Main Idea

Recall that if $a < 0$ and n is even, then $\sqrt[n]{a}$ is not a real number. Numbers like $\sqrt{-1}$ and $\sqrt[4]{-16}$ are not real numbers but are part of a number system called the complex numbers. Complex numbers have the form $a + bi$, where a and b are real numbers and i is called the imaginary unit with $i^2 = -1$ or $i = \sqrt{-1}$. The complex number is said to have real part a and imaginary part b. Powers of i are: $i^0 = 1$, $i^2 = -1$, $i^3 = -1$, and $i^4 = 1$. Two complex numbers are equal of and only if their real parts are equal and their imaginary parts are equal. Square roots of negative numbers can be defined as $\sqrt{-a} = \sqrt{(-1)a} = i\sqrt{a}$ if $a > 0$.

28. State the real part and the imaginary part of each of the following complex numbers:

a. $-4 + 2i$ b. $8i$ c. 12 d. $\sqrt{2} - 7i$

---

29. Compute the following:

a. i^5 b. i^0 c. i^{17} d. i^{18}

---

30. Simplify the following roots:

a. $\sqrt{-5}$ b. $\sqrt{-16}$ c. $\sqrt{-128}$ d. $-12+\sqrt{-12}$

---

31. If $a+2i=3-bi$, what are a and b?

Objective 8. Add and subtract complex numbers

Main Idea

To add (or subtract) two complex numbers, add (or subtract) their real parts and add (or subtract) their imaginary parts.

$$(a+bi)+(c+di)=(a+c)+(b+d)i$$

$$(a+bi)-(c+di)=(a-c)+(b-d)i$$

In the following problems, add or subtract, as indicated, and simplify.

32. $(8+2i)+(-3-4i)$

---

33. $(-3-12i)-(9+6i)$

---

34. $(4+i)+\left(2-\sqrt{-4}\right)-(17+8i)$

Objective 9.   Multiply and divide complex numbers

Main Idea

To multiply two complex numbers, use binomial multiplication and $i^2=-1$.
$(a+bi)(c+di)=ac-bd+(ad+bc)i$

To divide two complex numbers, multiply the numerator and denominator by the conjugate of the denominator. $\dfrac{a+bi}{c+di}=\dfrac{(a+bi)\cdot(c-di)}{(c+di)\cdot(c-di)}=\dfrac{(ac+bd)}{c^2+d^2}+\dfrac{(-ad+bc)}{c^2+d^2}$

In the following problems, perform the indicated operations and simplify.

35. $(3+4i)(5-2i)$

---

36. $\dfrac{1+3i}{5+2i}$

---

37. $i^{17}(3i+2)$

---

38. $\left(\dfrac{3+\sqrt{-16}}{2}\right)^2$

_____

39. $\dfrac{3+7i}{4i}$

_____

40. $\dfrac{2-\sqrt{-8}}{\sqrt{2}-\sqrt{-4}}$

_____

41. $\dfrac{(3-i)(7+2i)}{4-3i}$

_____

42. $\sqrt{-4}\sqrt{-3}\sqrt{-24}$

_____

Objective 1.   Use properties of exponents with real numbers

Main Idea

The rules of exponents can also be used with any real number as the exponent.

Product Rule of Exponents: $a^m \cdot a^n = a^{m+n}$

Power Rule of Exponents: $\left(a^m\right)^n = a^{mn}$

Power Rule for Products: $(ab)^m = a^m b^m$

Power Rule for Quotients: $\left(\dfrac{a}{b}\right)^m = \dfrac{a^m}{b^m}$

Zero Exponent: For any nonzero real number a, we define $a^0 = 1$. We leave 0^0 undefined.

Negative Exponents: $a^{-1} = \dfrac{1}{a}$ and $a^{-n} = \dfrac{1}{a^n}$ for nonzero a.

For a positive integer n, we define $a^{1/n} = \sqrt[n]{a}$ if $\sqrt[n]{a}$ exists.

In the following problems, perform the indicated operation and simplify so that only positive exponents remain.

1. $y^{1/4} \cdot y^{1/2}$

_____

2. $x^{-2/3} \cdot x^2$

_____

3. $\dfrac{x^{1/3}}{x^{-2/3}}$

_____

4. $\left(x^{2/3}\right)^{3/4}$

---

5. $\left(x^{-2/3}\right)^{3/5}$

---

6. $\left(4x^{4/3}y^{8/9}\right)^{3/2}$

---

7. $\left(\dfrac{2b^{1/3}}{6c^{2/3}}\right)^{3}$

Objective 2.   Simplify expressions involving rational exponents

Main Idea We can use the rules of exponents to simplify expressions involving rational exponents.

In the following problems, perform the indicated operation and simplify so that only positive exponents remain.

8. $(8x)^{1/3}(4x)^{1/2}$

---

9. $\left[\dfrac{9x^{2/3}}{16y^{5/6}}\right]^{3/2}$

---

10. $\left(x^{4/9}y^{2/3}z^{-1/3}\right)^{3/2}$

---

11. $\left[\dfrac{-8x^9y^{1/2}}{27z^{3/2}}\right]^{4/3}$

Objective 3. Convert between standard notation and scientific notation

Main Idea

Scientific notation is a way to write very large (positive or negative) numbers or numbers close to 0. Numbers in scientific notation have the form $N\times10^p$ where $1\le N<10$ and p is an integer. For example, the scientific notation of 2,654,000 is 2.654×10^6. A number written in scientific notation can be converted to standard notation by multiplying (when the exponent on 10 is positive) and dividing (when the exponent on 10 is negative). For example, we convert 7.983×10^5 by multiplying 7.983 by 10^5:

$$7.983\times10^5 = 7.983\times100,000 = 798,300$$

We convert 4.563×10^{-7} by dividing 4.563 by 10^7:

$$4.563\times10^{-7} = \frac{4.563}{10,000,000} = 0.0000004563$$

In problems 12-14, write the numbers in scientific notation:

12. 3,789,000

13. 49,000,000,000

---

14.   0.0000495

15. 0.000000278

In problems 15-18, write the numbers in standard form.

16. 4.372×10^5 17. 7.91×10^6

18. 5.6294×10^{-4} 19. 6.3478×10^{-3}

Objective 4.   Simplify expressions involving scientific notation

Main Idea

Multiplying two numbers in scientific notation involves adding the powers of 10, and dividing them involves subtracting the powers of 10.

In problems 19-21, multiply or divide, as indicated, and write the result in scientific notation.

20. $\left(6.250 \times 10^7\right)\left(5.933 \times 10^{-2}\right)$

21. $\dfrac{2.961 \times 10^{-2}}{4.583 \times 10^{-4}}$

22. $\dfrac{7.983 \times 10^{-4}}{3.772 \times 10^6}$